Earth's Endangered Creatures

POLAR BEAR RESCUE

Written by
Jill Bailey

Illustrated by
John Green

This series is concerned with the world's endangered animals, the reasons why their numbers are diminishing, and the efforts being made to save them from extinction. The author has described these events through the eyes of fictional characters. Although the situations described are based on fact, the people and the events described are fictitious.

A Templar Book
First published in Great Britain in 1991
by Heinemann Children's Reference
A division of Heinemann Educational Books Ltd
Halley Court, Jordan Hill, Oxford OX2 8EJ
Devised and produced by The Templar Company plc
Pippbrook Mill, London Road, Dorking, Surrey RH4 1JE
Copyright © 1991 by The Templar Company plc
Illustrations copyright © 1991 by The Templar Company plc

Editor Andy Charman
Designer Mike Jolley
Colour separations by Positive Colour Ltd, Maldon, Essex
Printed and bound by L.E.G.O., Vicenza, Italy

British Library Cataloguing in Publication Data
Bailey, Jill
Polar bear rescue
1. Polar bears. Conservation
I. Title II. Green, John 1918 Sept. 3- 111. Series
639.97974446

ISBN 0-431-00112-X

CONTENTS

REALM OF THE POLAR BEAR

Anne-Marie Dupont pressed her nose to the window of the helicopter. They had left the trees behind, and were now flying over a world of ice and snow. It was March; the dark Arctic winter was over. A weak yellow sun caught the frozen surface, making the ice gleam and sparkle. Gusts of wind whipped up the powdery snow into a miniature blizzard.

Anne-Marie had been studying bears – grizzly bears – for several years now, but this was her first chance to study the largest bear of all, the polar bear. She had come to help George Weinberg, another Canadian biologist, who was trying to count the polar bears on the remote islands of Canada's Northwest Territories.

Polar bears grow fat on a diet of seals in winter. The males can stand up to 3.38 metres high at the shoulder, and may weigh 700 kilograms.

The helicopter followed the coast. The sea was still frozen, a flat expanse of ice covered with a dusting of snow. Here and there massive blocks of ice had been crumpled and pushed into huge pressure ridges. Dark water gleamed in long, narrow cracks in the ice. George and Anne-Marie scanned the white expanse.

"That water never freezes," remarked George. "It is kept ice-

In the far north, the sun sets for the last time in late November, and does not appear above the horizon until the following spring.

free by powerful sea currents. These 'leads' of water are good places for seals, so the polar bears come here to hunt. They spend the winter on the sea ice. It is the best time of year for hunting. Look, there's a bear."

The polar bear was almost invisible among the blocks of ice. Its white coat was certainly a wonderful camouflage. It was a fat, male bear with a thick coat, and its muscles seemed almost to flow along as it padded across the frozen sea. This was the animal they called "The King of the Arctic".

As she got used to the dazzling white landscape, Anne-Marie began to see more bears. They flew over a bear with two large cubs running after her.

"Those cubs look almost as big as their mother," remarked Anne-Marie.

"Polar bear cubs usually stay with their mother until they are at least two years old," said George.

The polar bear's coat blends with the snow and ice, allowing it to creep up on its prey unnoticed.

"It takes them a long time to learn how to hunt for themselves. In any case, that sea ice is quite thick. To get at the seals in their breathing holes the bears may need to smash through it. Those cubs are probably not strong enough to do that yet, so they would have trouble surviving the winter alone."

The land began to rise.

"That's Southampton Island," said George, "our first stop. Many female polar bears make their winter dens here. We shall count the dens, and tag the females to see where they go in the summer. We need to know which groups of bears use which denning areas, so that we can protect the most important areas."

A polar bear with her cub. Most polar bears have twins, but single cubs are not uncommon.

Above: an expedition prepares for a journey across the Arctic sea ice. They have to take all the food and fuel they will need right from the beginning.

Left: a polar Inuit ties protective boots on to his husky's feet which can be cut by thawing ice.

Right: skidoos are faster than a team of husky dogs and a sledge, but they can run out of fuel, and they do rely on their driver to spot crevices in the ice.

The helicopter landed near a small cabin at the foot of the cliffs. They unloaded their gear and made for the hut. Inside were two camp beds, a table, a lamp and some old packing cases that served as furniture. Anne-Marie lit the stove, and soon the welcome smell of soup filled the cabin. Meanwhile, George was busy outside.

"Supper's ready, come and eat," called Anne-Marie.

George came in, his cheeks pink from the cold. They had just finished the soup when there was a strange noise outside. As they got up to see what it was, there was a loud bang and a flash. George threw open the door just in time to see a large polar bear galloping away.

"I laid a trip-wire around the hut to set off the thunder-flash," he explained. "I was afraid the bears might smell the cooking."

The following morning, Anne-Marie and George set off in search of the dens. The skidoos – small, light snowmobiles – carried them easily across the ice and into a narrow valley. George pointed to the snow-clad hillsides.

"The bears like to den in deep snowdrifts," he said. "The snow insulates them from the cold. Here, they are also well away from the big males hunting on the sea ice. Male polar bears will catch and eat small cubs if they can."

After a couple of hours, George suddenly stopped.

"Look up there," he said, and he pointed to a small, dark spot high up on the hillside. "A den!"

Anne-Marie and George left the skidoos halfway up the slope, so as not to disturb the bear, and cautiously climbed up to the den. George had a loaded gun ready in case the bear attacked them. There were paw-marks around the entrance.

"Good," said George. "That means it's occupied. Let's wait until the bear emerges and see how many cubs she has."

They moved a safe distance away downwind from the den, in case the bear smelled them. Then they built a wall of snow blocks to hide behind, got out their binoculars and waited. Even though the wall kept off the wind, it was very cold, well below freezing.

After three miserable hours, a black nose appeared in the den entrance. A bear pushed her head out and sniffed the air. Slowly she pulled the rest of her body out, stiff after the long winter in the den. She stretched, then rolled on her back and rubbed her body in the snow. She began to walk down the hill, slipping with every step, until eventually she gave up and slid down on her belly, rolling and

Polar bears shelter from the wind in snow hollows. They don't mind the snow covering them, because it keeps out the wind.

Left: the mother polar bear cannot go far from her cubs until they are strong enough to follow her. Rolling in the snow helps to keep her coat clean.

grunting as if she was thoroughly enjoying herself.

"She's got very long legs," whispered Anne-Marie.

"They look long because her belly is so thin," said George. "That bear hasn't eaten since she entered the den in October. She's been living on her fat all winter and must be very hungry."

The polar bear was almost out of sight down the slope when Anne-Marie and George heard a high-pitched bleating sound. Two tiny cubs peered out of the den entrance. This was their first glimpse of the world outside. Gingerly, one cub put a paw forward to test the snow, then withdrew it, not sure what to make

A polar bear cub takes his first look at the Arctic world.

of it. Their mother came hurrying back, and led them for a short walk. Their tiny legs were very wobbly, for they had not walked more than a few metres until now.

"They look quite big," whispered Anne-Marie.

"They are about two and a half months old," said George. "Newborn polar bear cubs are very tiny, about the size of a large rat. They weigh only about 700 grams, and are blind and deaf, with just a thin coat of woolly fur. After about a month their eyes open, and about ten days later they can see. They don't start to walk until they are about seven weeks old."

After about half an hour the mother bear led her cubs back into the den, and Anne-Marie and

George went back to the cabin.

About ten days later, on their way to visit some other dens, George and Anne-Marie saw the mother bear leading the cubs towards the sea ice.

"She must be very hungry now," commented George. "It's dangerous for the cubs out there on the ice, but their mother needs to start hunting seals again."

The mother polar bear needs to go on to the sea ice to hunt. It is dangerous here for the cubs. The large male bears may consider them a tasty meal.

"Let's go and look at the den," suggested George.

They wriggled inside the den. There was a narrow entrance tunnel about 3 metres long, which sloped up to the main chamber.

"Warm air rises," said George. "By sloping the tunnel, the bear makes sure the warm air inside does not escape. The Inuit use the same trick when building their snowhouses or igloos."

The main chamber was just big enough for a bear to turn around in. The ceiling of the chamber was covered in scratch marks.

"The bear likes to keep the snow above reasonably thin, so that air can get through," said George.

The snow den provides a warm nursery for the cubs, who are born in the middle of winter. They feed on their mother's milk while she snoozes.

"She also covers the cubs' droppings with snow to keep the floor clean."

"It's warm in here," said Anne-Marie.

"The bear's body heat is trapped by the snow," explained George. "The air in the den can be up to 35 degrees warmer than outside. Let's go and check the other dens. Tomorrow we have to leave early."

The helicopter returned the following day, and they flew out to the sea ice, where George wanted to tag some bears. Anne-Marie admired the skill of the helicopter pilot. When they spotted a female bear, he flew low beside her as she fled, until George could get close enough to fire a tranquillizing dart from his rifle. The bear started as the dart struck home, then galloped away. Soon she was staggering, and within a few minutes she had collapsed on the ground in a deep sleep.

Polar bears' fur consists of two layers; long, stiff hairs for protection and a soft, woolly underfur for keeping warm.

Anne-Marie cradled the polar bear's massive head in her lap as George pulled out a small tooth. He wanted to count the growth rings on it to find out the bear's age. Anne-Marie could feel the hot breath on her hand as the bear sighed deeply in her sleep. George

Here, scientists weigh a tranquillized bear in Churchill. The bear may weigh half a tonne.

injected an antibiotic to make sure the bleeding gum did not get infected. Then he measured the bear with a tape.

He beckoned to the helicopter pilot, who was keeping watch with his gun. The pilot ran to fetch a large canvas sling. Together they heaved the sleeping bear on to it, then secured the sling to the helicopter. The pilot started up the helicopter and lifted the bear off the ground. Then Anne-Marie could see the weighing scale attached to the helicopter.

The helicopter laid the bear carefully back on the ground, and George rolled her over so that she would be comfortable when she woke up. Then Anne-Marie helped him fix a radio collar around her neck and clip a plastic number tag in her ear.

"The collar is designed to fall off after about two years," said George. "The tag will help us to recognize her after that."

Suddenly the helicopter pilot shouted a warning. Anne-Marie looked up. There, behind George, only about 100 metres away, was a huge male polar bear.

While the bear slept, George and Anne-Marie fitted the radio collar. This would send out signals for about two years, allowing them to plot her movements.

"Don't move," whispered George, "and whatever you do, don't look him in the eye."

There was a whir of propellers as the pilot started up the helicopter. The bear jumped backwards, stared defiantly at the helicopter and then, as it swerved towards him, lumbered off.

"Let's get finished quickly, before he comes back," said George.

He injected the tranquillized bear with another chemical to wake her

Captured bears are often numbered. They can be identified easily when they are recaptured and changes in weight and height recorded.

up, and they climbed back into the helicopter.

"This is the bears' mating season," said George. "This bear has no cubs with her, so she is ready to mate again. In the autumn her radio signals will lead us to her denning area."

As they moved off, the large male appeared over the horizon. He was determined not to lose sight of the female.

"He may stay with her for several days," George continued.

"How did he find a mate in all this space?" asked Anne-Marie.

"When a female is ready to mate," said George, "her urine has a special smell. Polar bears have a very good sense of smell, and she will probably attract several males. They may even fight over her."

In the distance, they could see the great male was nuzzling the slowly awakening female.

In several weeks, George and his assistants tagged about 20 bears, and fitted radio collars to two more single females. It was time to move to their summer cabin on Devon Island.

The male bear is not put off by his mate's radio collar. The collars are so powerful that their signals can be picked up by satellites and followed hundreds of kilometres away.

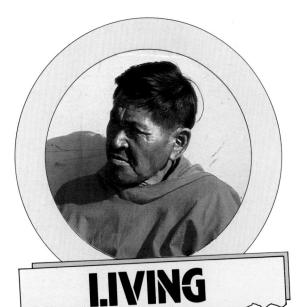

LIVING
WITH BEARS

Sam Ipalook felt a tingle of excitement as the plane climbed over the wind-shaped trees around Churchill and headed north. Although his family had given up their nomadic life and settled in modern cabins in Churchill when he was a boy, he still had the Inuit's instinctive love of the open spaces of the Arctic north. Sam worked for the Canadian Broadcasting Corporation (CBC) as a presenter and producer of radio programmes, based in Churchill.

CBC had decided that a feature on what the polar bears were getting up to in summer would liven up their news broadcasts. Sam was going to join George and Anne-Marie, who were watching polar bears on Devon Island, far to the north.

As they crossed the sea ice surrounding Devon Island a strange sight loomed ahead. Steep islands of limestone rose above the flat, ice-covered ground, their sides rimmed with cliffs that dropped hundreds of metres to the beach below. No wonder this was such a good place to watch polar bears. If people were on top of one of these rocky "islands" the bears could neither see nor smell them. The bears would continue to behave perfectly naturally.

They landed on the largest island, Caswall Tower, near the scientists' hut. Anne-Marie and George were sitting by the edge of the cliff scanning the land below with large telescopes. Sam went to join them.

Left: the ice was beginning to melt. Pools of water lay on the surface and cracks appeared. The bears were now hunting far out on the ice.

Above: these trees are often called "flags" because they have branches on one side only. The wind blows so hard that branches cannot grow on the windward side.

Sam could feel the magic of the ice all around him. The silence – he had forgotten what real silence was like. The only sounds were the occasional creaking and cracking of the ice as it broke away from the shore, and the calls of snow geese and terns as they flew overhead.

"The bears are busy this morning, Sam," said George. "Come and look."

The polar bear has fur on the soles of its feet to prevent them freezing on the ice. It has long sharp claws and very tough skin on its pads, which grip the slippery ice.

Sam settled himself down at a telescope. The ice was beginning to melt. Pools of water lay gleaming in the sun. The snow had already gone, and the surface of the ice was wet and slippery. Seals were basking in the sun beside their breathing holes. He could see a large polar bear stalking a seal. It was creeping forward stealthily, rather like a cat stalking a bird. Every time the seal looked up the bear froze. When it got to within a few metres, it pounced in one enormous leap. The seal was too quick for it, and dived into the water.

"That must be an old seal," remarked Sam. "It has learned to react quickly. The young ones tend to freeze in horror when they see a polar bear."

"I'm surprised the seal doesn't see that black nose coming," laughed Anne-Marie.

"Seals are short-sighted," said Sam. "In fact, the bear has black skin all over, under its white fur. The black skin helps to keep it warm by absorbing heat."

The ringed seal is the polar bear's main source of food. The seals live on the sea ice, and hunt for food in the cold water below.

Later that morning, Sam spotted a polar bear feeding on a baby seal.

"Where did that come from?" asked Anne-Marie.

"The seals give birth in dens under the ice," explained Sam. "The bears can smell them even through the ice, and they are strong enough to smash the ice to reach them."

"Our studies show that the most successful way for a bear to hunt is to lie in wait at a breathing hole," said George. "Seals have very good hearing. They can hear anything walking on the ice above,

Ringed seals spend much of their time basking in the sunshine. They have to look out for bears.

so the bear must keep very still. A tiny stream of bubbles rises to the surface just before the seal surfaces. This warns the bear to get ready to pounce."

"We used to hunt seals the same way," said Sam.

The following day they saw a mother with young cubs on the ice. The cubs were chasing each other and must have been alerting every seal in the area.

"Poor bear!" said Anne-Marie. "She'll be lucky to catch a meal with all the noise they're making."

Anne-Marie was wrong. Later that day, the mother bear spotted a young male bear with a freshly killed seal. She was much larger than the male. As she charged towards him, he dropped his dinner and fled. The bear family settled down to feed. At first, they ate only the skin and the blubber, or fat.

"If they get chased off by an even bigger bear, at least they will have eaten the best bits," chuckled Sam.

When food is scarce, there is competition among bears. Young bears often lose their kill to older ones.

It was autumn before Sam had another opportunity to visit the bears. Or rather, the bears visited him. Churchill lies on the west coast of Hudson Bay, in Canada. Here, the sea ice freezes early, and polar bears from near and far come to Churchill to wait for the ice to freeze over.

In summer, when the sea ice melts, they are forced to come ashore, where there is very little food for them. They may catch the occasional duck, diving and taking it from below by surprise. They may also catch lemmings, tiny voles that live under the snow, but these are hardly a mouthful for a large bear. Some bears even eat seaweed or berries. Most just lie around and sleep, living off their fat and saving their energy.

The town is a great temptation for the bears. Smells of cooking and food scraps are very appealing to a bear that has not eaten for several months. The outlying cabins suffer most. The bears break through windows and even walls to get to the food, and most families keep a loaded gun at the ready. Fortunately, very few bears are shot because the town has a bear patrol on duty day and night while the bears are in town.

Sam was to join the bear patrol to report on the invasion. Touring the outskirts of the town in their vans, the patrol used firecrackers fired from a long-range rifle to scare away the bears.

Signs are put up around Churchill in autumn warning people to keep away from the area where the bears like to sleep.

When the sea ice melts, the seals hunt out at sea. The polar bears take to eating berries, seaweed, lichen and even mushrooms.

The bears visit the rubbish dump at Churchill for scraps of food. It is dangerous for both bears and people if they stay.

Sam found the patrol at the rubbish dump on the edge of town. This was a favourite place with the bears, who were attracted by the smell. They rummaged in the dump, licking out tins and jars, and fighting over scraps.

The dump was also popular with tourists, an important source of income for Churchill. A party of tourists was watching the bears from the safety of a large minibus, their cameras sticking out through the open windows.

Tourists visit Churchill to see the polar bears. This creates jobs and income for the residents of the town.

Some of the bears had large black numbers painted on them. They had also been tagged by Anne-Marie and George. The paint would be shed with the fur when the bears moulted.

Suddenly, one of the bears reared up on its hind legs to peer inside the minibus, and tried to snatch a camera. The driver of the minibus sounded his horn loudly and drove off at speed.

"That's number 27," said the bear patrol chief, Douggie. "He's a trouble-maker. He raided some cabins last week, so we picked him up and flew him 100 kilometres up the coast. I guess this time we'll just have to put him in the pound."

Later that day, they followed number 27 back to his sleeping hollow among the willow bushes. They set a trap nearby. This consisted of some seal meat inside a large metal drum. A trap door shuts the bear inside when it takes the meat. Once the bear was inside they would take him to the bear pound, where he would have to stay until the sea ice froze.

Some bears are such a nuisance that they have to be transported out of town by truck and then helicopter.

Anne-Marie and George were watching polar bears from a tundra buggy at Cape Churchill, about 65 kilometres from Churchill. Anne-Marie was surprised to see such numbers of bears – they are usually rather solitary animals. Most of the bears were males. The pregnant females were already safe in their dens, and the mothers with small cubs were too nervous to go there.

As she watched, two bears began to bite at each other's faces. They reared up on their hind legs and started to wrestle.

"Don't worry," said George, "they're not fighting – that's play wrestling. They won't hurt each

Despite their great size and weight, polar bears can walk on ice too thin to support a human.

other, even though one is bigger than the other. They only fight seriously in the spring, when they are competing for the females. By testing their strength in this way, bears learn who is the strongest and so avoid serious injury later."

Anne-Marie was thrilled. She had never seen bears playing like this.

Some bears preferred to save their energy, and spent their time dozing in hollows they had scooped out of the snow among the willow bushes. Churchill needs its warning signs – it is easy to stumble over a sleeping bear that is covered in drifting snow, and bears do not like being disturbed.

By late October the ice was beginning to form. As the ice thickened, impatient bears tried it out, splaying out their legs to spread their weight. Often they crashed through and swam sulkily back to the shore. Eventually, the ice was thick enough to support the bears' weight and they left Churchill to hunt seals again.

Polar bears test their strength by play-fighting. Eventually, the weaker bear will run away. Neither is seriously hurt.

THE ARCTIC HUNTER

Inuterssuaq and his wife Elispee were just putting the finishing touches to the igloo when Sam Ipalook and the film crew arrived on their skidoos. They were going to make a film about the old Inuit way of life. Here at Upavik, on the west coast of Greenland, a few families still lived in the traditional way, by hunting seals and other wild animals. Inuterssuaq did not like to depend upon supermarkets and government hand-outs. He preferred to be independent.

Usually the family lived in a small wooden hut. Occasionally Inuterssuaq would make an igloo if he was out hunting far from home. He was making this igloo specially for the film. The barking of the husky dogs announced the arrival of Sam and his friends.

Inuterssuaq had one modern luxury, a skidoo for hunting in winter. In summer, when the ice was melting and cracks could appear almost anywhere, he preferred to use the dogs. Unlike the skidoo, dogs could sense danger and avoid it.

Inuterssuaq was looking forward to taking the crew on a polar bear hunt. This was the highlight of his year. Families that live by hunting must have a licence and are allowed to take two bears a year. For the Inuit, polar bear hunting is something very special. An Inuit boy becomes a man the day he kills his first polar bear.

Left: an Inuit fur trapper. Many Inuit still live by hunting. This man has caught a marten, an animal related to otters and badgers.

Igloos are surprisingly warm inside. Few Inuit live in igloos today, but they sometimes build them as shelters while out hunting.

Sam had abandoned his town clothes for the traditional Inuit dress of caribou fur anorak, sealskin boots and polar bear fur trousers, just like Inuterssuaq's. These clothes would keep him warm out on the ice. Even in summer, a blizzard could blow up at any time. The bearskin trousers were his prized possession; it took a whole bear skin to make them, but they were wonderfully warm. Elispee had used the long fur from a polar bear's legs to trim her boots.

The dogs were straining impatiently at their leads. They had noticed the hunting preparations. Inuterssuaq produced a piece of polar bear fur, spat on it, then rubbed the runners of the sledge with it. Instantly, the thin film of water froze to form a slippery coating of ice.

They set off under a blue sky and brilliant sun, racing over the ice towards the hazy horizon. Seals plopped into the water as they approached. Inuterssuaq talked about the polar bears.

"We use almost all parts of the bear," he said, "the skins for clothes, the meat for the dogs. Polar bears suffer badly from worms, and these can infect humans, too, so we don't eat their meat ourselves. If we don't need

The equipment was loaded on to three wooden sleds. Each would be pulled by a team of up to ten dogs. It was some time before they were ready to go.

the skin, we sell it – it fetches about £300 a foot, measuring from head to tail."

After about 10 kilometres, Inuterssuaq stopped and got out a long telescope to scan the horizon for polar bears.

Although the Inuit traditionally wear clothes of real fur, researchers and explorers wear polar suits of man-made fibres which are just as good at keeping out the cold.

Sam asked Inuterssuaq how far the bears roamed.

"The polar Inuit name for the polar bear is 'pisugtooq', which means 'the great wanderer'," replied Inuterssuaq. "Most bears don't travel more than about 100 kilometres from their place of birth, but some of the bears of east Greenland regularly cross the ice to Svalbard, off the north coast of Norway, 3,000 kilometres away."

"I didn't know they lived in Norway, too," said the cameraman.

"People think they once lived even in southern Norway, and in southern Alaska too," said Inuterssuaq, "but the hunters further south killed all the bears there. Today they still live in the USSR, Svalbard, Greenland, Canada and Alaska. There are about 12 distinct

---

The Inuit have always hunted polar bears, but they take only as many as they need. Polar bears and Inuit have lived and survived together for thousands of years.

populations altogether, in 13 million square kilometres of snow and ice."

"Did Indian hunters kill off the southern bears?" asked one of the camermen.

"No," replied Sam. "It was the Europeans. First they came for the whales. Then, when there were no more whales left, they hunted the seals until they, too, began to decline. So they started on the bears, not killing them for food but for the skins, to sell them in Europe, and just for 'sport'. Then in the 1940s hunters began to come in by the plane-load. Mostly rich Americans. They even chased the bears in planes."

Inuterssuaq turned the sledge to the west, and soon the cries of gulls led them to traces of a polar bear kill.

Polar bears are good swimmers. They swim with a 'doggy-paddle' stroke and head stretched forward. They have been seen swimming over 300 kilometres from land.

A rather messy seal carcass lay on the ice, and an Arctic fox scampered away as they approached.

"I'm surprised the bears can keep such white fur," remarked Sam.

"They roll in the snow to clean themselves after eating," said Inuterssuaq. "They also lick their fur, and perhaps even wash if there is water nearby."

"Did you notice the fox?" asked Inuterssuaq. "They survive the winter here almost entirely by feeding on the remains of polar bear kills. Each fox follows its own polar bear, waiting for a kill. The bears get rather fed up with them, but the foxes are too quick for them to do much about it."

They followed the bear's tracks in the snow.

"This is a male bear," said Inuterssuaq. "His footprints are about three times larger than the female's, but then he is about three times bigger than her."

Now they could see the bear in the distance. It had already got wind of them. It stood stock still, one paw raised in the air. Then it reared up on its hind legs and sniffed the air, snorting in annoyance.

The film crew had their cameras ready. Inuterssuaq unleashed the dogs, who charged towards the

bear. At first the bear stood its ground. Then, as the dogs got closer, it slowly backed away. Inuterssuaq raised his rifle.

Arctic foxes are a familiar sight wherever there are bears. They follow the bears hoping to snatch a scrap of food.

Left: polar bears have an amazingly powerful sense of smell. They often stand on their hind legs to sniff the air, and to get a better view of the scene.

Suddenly, the bear slipped sideways into a lead of dark water they had not noticed, and swam quietly away.

"It's getting too late to chase another bear," said Inuterssuaq. "We shall have to try again tomorrow."

They sat down to enjoy steaming mugs of coffee before returning to the village. Sam was enthralled by the beauty of the scene, and the incredible silence.

"I had forgotten how quiet the Arctic could be," he said.

"It's not so quiet these days," said Inuterssuaq. "There are oil exploration rigs, ice-breakers and planes. There are too many people here; one day there will be a big disaster, you will see. If one of those oil rigs leaks, the oil will spread for a great distance under the ice, and that is where the shrimps live that the fish feed on. The seals eat the fish, and the bears eat the seals – in time they will all suffer."

"There's an even bigger threat," said Sam. "It's called the 'greenhouse effect'. All the gases industry is putting into the atmosphere, the burning of petrol by cars, and the carbon dioxide from the burning of the rainforests, are causing the planet to warm up. Already parts of the Arctic ice sheet have thinned by one third. If it melts, the polar bear will lose vast areas of its home. Unless we can stop this pollution, that beautiful sun we are watching may destroy the world of the polar bear."

Left: an oil rig in the Arctic sea. Increased industry means more people and this means that wildlife is disturbed. Man and bear are meeting more and more often.

It is too soon to say that the polar bear has been saved from extinction, but for the moment, it is still "The King of the Arctic".

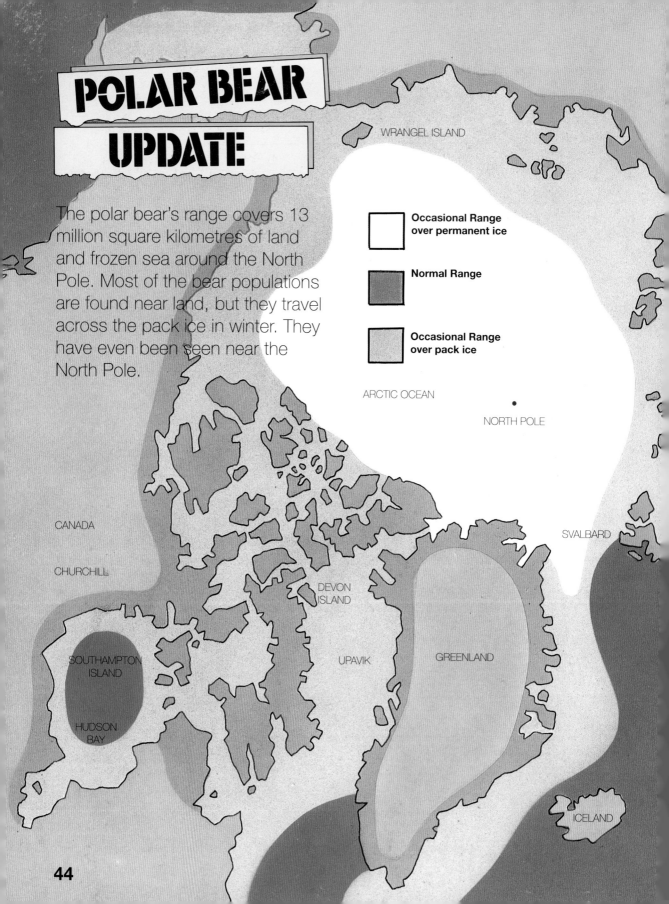

POLAR BEAR UPDATE

The polar bear's range covers 13 million square kilometres of land and frozen sea around the North Pole. Most of the bear populations are found near land, but they travel across the pack ice in winter. They have even been seen near the North Pole.

☐ Occasional Range over permanent ice

■ Normal Range

▨ Occasional Range over pack ice

WRANGEL ISLAND

ARCTIC OCEAN

• NORTH POLE

SVALBARD

CANADA

CHURCHILL

DEVON ISLAND

UPAVIK

GREENLAND

SOUTHAMPTON ISLAND

HUDSON BAY

ICELAND

44

A conservation success story

Soon after the Europeans discovered the Arctic, they began to hunt and slaughter the large animals that lived there. First they hunted whales, then seals and, finally, bears.

In the 1960s, the Russians became worried about the slaughter. They had banned the hunting of polar bears in the Soviet Union in the 1950s. The New York Times also ran a feature on the unsporting slaughter of polar bears. The five polar nations – the United States, Canada, Denmark (which governs Greenland), Norway and the Soviet Union, got together to make plans to study the polar bears, and in 1973 they signed an agreement to protect the bears. The International Union for the Conservation of Nature and Natural Resources (IUCN) set up a special research group to advise on managing the bears. It was thought that there might be only 5,000–10,000 polar bears left in the entire world.

● In most countries today, hunting of polar bears is banned, except by native people, such as the Inuit, who live by hunting. In recent years, guns have made these people's hunting much more successful. Now they are only allowed to hunt a certain number of bears. This ensures that the bear population is not seriously affected. They are not allowed to hunt during the time of year when female bears are pregnant and they are forbidden to shoot females with cubs. Since these rules were brought in the polar bear population has increased to 20,000–40,000.

● Some countries have set aside special reserves where polar bears are completely protected. Canada has its Polar Bear Pass, a popular migration route for the bears, Denmark has the East Greenland National Park, and the Soviet Union has Wrangel Island, off the coast of Siberia.

● The most important remaining threat to the bears is the warming of the planet due to the "greenhouse effect". It will be much harder to introduce controls of pollution worldwide to prevent the polar ice sheets melting.

INDEX